IMPRINT
CLASSICS

Horrie's Miscellany

The Story of a Good Dog

Introduced by Tim Bowden

ETT IMPRINT
Exile Bay

ETT IMPRINT
PO Box 1906
Royal Exchange NSW 1225
Australia

Copyright © Idriess Enterprises 2024

First electronic edition published by ETT Imprint 2024

ISBN   978-1-923205-46-8
ISBN   978-1-923205-47-5

Compiled and designed by Tom Thompson from the Idriess and Moody Archives, with special thanks to the Moody family

Cover: Horrie with his master Private James Moody, during Horrie's Middle East Campaign 1941.

for Lee & Ian Moody
and Betty Featherstone,
a great friend of the Moodys
whose husband Feathers'
corporal stripes are on Horrie's uniform

Introduction

Tim Bowden

I have the honour of introducing this unlikely pictorial record of a small mongrel dog, who rejoiced in the name of Horrie The Wog-Dog so named by the Australian soldiers who adopted him at the beginning of World War II in the Western Desert in Egypt in 1941. The little white pup was noticed by Private J B Moody of the 6th Division AIF who was training for whatever the future might hold.

Moody's mate, Don Gill, attached to their Signals Platoon as a dispatch rider was navigating the desert on his motorcycle, following a compass course. Moody, who had started 15 minutes before Gill, was waiting for him, smoking a cigarette near the "Wog's" bakehouse, a rough shed where the bread (and sand) were baked for the troops by Arabs under the supervision of Tommy soldiers. Hearing the sound of the approaching motorcycle, Moody's attention was attracted to a small white pup racing from rock to rock, a grim earnestness in his obvious tiring movements chasing lizards he was failing to catch. 'What a comical little joker', laughed Don. His coat was a dusty white, emphasised by a sandy coloured stripe running along his back. On quaint stubby legs he stood barely a foot high, his front legs bowed like those of a miniature bull-dog. His long body was out of proportion to his height. His extraordinarily intelligent face was pinched a forlorn, with an expression now changing from dire suspicion to one of inquiring hope. His stub end of a tail rose erect, his sharp little ears alternately stood to attention, then dropped at ease.

The two Australians decided to help him by joining in the quest for lizards – without success. 'Poor little pup', sympathised Don, and patted his head. 'You are an outcast from away home'. They decided to adopt him, despite no pets being allowed in the camp, where he was quickly accepted as a mascot by the soldiers. But what to call him? He was an Arab dog, so someone suggested 'Wog Dog'. 'Wog' was the Aussie soldiers nickname for the Arabs. The Arabic meaning of Wog, was 'Worthy Oriental gentleman'. Another suggestion was to call him Horrie, 'A good old Australian name'. Horrie The Wog-dog, it was decided, 'Fitted the pup like a glove'.

So began the career of Horrie whose adventures lasted through the rest of the war, now splendidly detailed in Ion Idriess's book, *Horrie The Wog-Dog* working from Moody's diaries and notes.

However before the war ended, getting pets back to Australia was forbidden. One very loved cat, another mascot for a different unit, was detected on the ship taking the troops home from the Middle East one day before reaching Fremantle in 1944. It was unceremoniously thrown overboard by the powers that be. Horrie was safely smuggled ashore, but the records show that under Quarantine Regulations, Horrie was destroyed on 12 March 1945. Well so wrote Ion Idriess, but if you read to the end of *Horrie's Miscellany* you may get an unexpected surprise…

Finding Horrie

James Moody

Don Gill and myself were the first to find Horrie and it came about in the following manner. Our jobs in the platoon were motor cycle dispatch riders, when not actually employed on a dispatch run we spent the days in maintenance and repairs to the cycles, after tinkering about with the engines we would often give the cycle a try out across the rocky and sandy desert that made up the Ikingi Maruit area. Ikingi Maruit was on the fringe of the Western Desert and some twenty odd miles from Alex.

As the going was fairly rough for cycles we knocked out a fair bit of them and acquired a little experience in rough riding that later stood us in very good stead. We evolved a little game to make these trips across the sandy waste more interesting, the idea being to ride on a compass bearing for a certain distance, then another bearing was taken until a trip of five or six legs of various bearings was completed. We started the first leg from a point known as the Wogs Bakehouse, it was a rough shed arrangement where the bread (and sand) was baked for the troops by Arabs under the supervision of Tommy soldiers.

Before starting we agreed upon a bearing and distance for the first leg, then one rider would away about half an hour before the following rider. After completing the leg the first rider would then select another bearing and distance and write it down on a piece of paper, place it on the ground and cover it with a small heap of stones, making the heap about twelve inches high, he would then complete the second leg and repeat the procedure until five or six legs were completed. The task of the following rider was to complete the first leg, locate the heap of stones and so get the clue for the second leg. It was fairly difficult as one had to ride exactly on the bearing otherwise difficulty was experienced in finding the small mound of stones, to keep on the bearing all sorts of obstacles were encountered such as deep

VX13091

MEDICAL EXAMINATION

I have made full and careful examination of the abovenamed person in accordance with the instructions contained in the Standing Orders for Australian Army Medical Services. In my opinion he is—*

1. Fit for Class I. _as per D.' Harvey_ Capt.
~~2. Temporarily unfit for Class II~~ A.O. 14th Area
~~3. Unfit for military service†~~ 28/3/40

Place _Prahran_   Date

Signature of Examining Medical Officer

* Classifications which are inapplicable to be struck out.   † Reasons for unfitness to be stated.

OATH OF ENLISTMENT‡

I,   JAMES BELL MOODY   swear that I will well and truly serve our Sovereign Lord, the King, in the Military Forces of the Commonwealth of Australia until the cessation of the present time of war and twelve months thereafter or until sooner lawfully discharged, dismissed, or removed, and that I will resist His Majesty's enemies and cause His Majesty's peace to be kept and maintained, and that I will in all matters appertaining to my service faithfully discharge my duty according to law.

So Help Me God.

Signature of Person Enlisted   _J.B Moody_

Subscribed at   PRAHRAN   in the State of   VICTORIA

this   _19th_   day of   _March_   1940

Before me—

Signature of Attesting Officer   _Harvey_ Capt.
A.O. 14th Area

‡ Persons who object to take an oath may make an affirmation in accordance with the Third Schedule of the Defence Act. In such case the above form will be amended accordingly and initialled by the Attesting Officer.

Wilke and Co. Pty Ltd., Printers, 19-47 Jeffcott Street, Melbourne.

shifting sands, and in parts rocky surfaces, these spots had to be negotiated without deviating far from the bearing.

On this particular day I was first away and having completed the last leg was waiting for Don to catch up. Propping the cycle up I sat in the small shade it offered and dreamingly smoked a cigarette. My attention was attracted to a tiny white object darting across my line of sight. Coming to life I stood up and watched a small pup racing from rock to rock. He did not appear to notice my presence as his entire interest was in his task of trying to catch small lizards that darted from rock to rock in order to escape him. Amused I stood quietly watching him when the noise of Don's approaching cycle attracted his attention. His alert little ears were pricked to catch the sound and he watched Don approach me with an enquiring look. As Don stopped and dismounted the pup withdrew a little distance and continued to cautiously watch us. I drew Don's attention to the pup and remarked "I wonder what that little bloke is doing way out here" the last leg had taken us well out into the sandy waste, there being nothing but sand and rock for some fifteen odd miles in circumference. "Looks like he's blown through from somewhere" I answered. ('blown through' is a term used by the troops meaning absent without leave) "He's a funny looking little joker" exclaimed Don.

The little pup by this time had decided we were not as important as the lizards and had resumed his self-appointed task of ridding the Egyptian desert of lizards. He certainly was a funny looking little joker, he was all white except for a broad sandy coloured stripe along his back, he stood about twelve inches high on very short little legs, the front two legs were similar in shape to that of a bull dog, the length of his body was all out of proportion to his height. In fact 'Longfellow' was suggested as a name for him by Feathers, his sharp little ears alternately stood to attention and stood easy and he possessed a tiny little stub tail that stood as erect as a mast.

He was a bit doubtful about us at first and our efforts to get him to come to us were unsuccessful, as we called him he would

Private James Moody, 2/1 Machine Gun Company.

view us with his little head on one side then turn his attention once again to the lizards. "Do you reckon he's an Arab dog" I asked. "Could be anything I guess" replied Don, "He certainly doesn't seem to understand our lingo" 'Here pup' 'Good dog' 'Here boy' etc meant nothing to the little pup, although when we called him he would watch us cautiously, he did not seem exactly afraid.

Eventually we won his confidence by turning rocks over and sending the lizards that hid underneath scurrying in all directions. The pups little stub tail betrayed a good view of our effort to help him, he would select one particular lizard, however the chosen one would invariably escape him and seek shelter under another rock. The little pup would look up at us with his beautiful brown eyes appealing to us for further assistance. 'This is not a game, I'm hungry' they seemed to say. I eventually picked him up and I could feel his little ribs only just covered by his silky soft short haired coat.

"Doesn't seem to be doing so good" I remarked. Don patted his head "Poor little pup, only a puppy too" We guessed his age to be about six months. "We might feed him up a bit back at camp" I suggested. "Good idea" replied Don.

By this time the pup had gained complete confidence, and resting his chin on my arm he closed his eyes. The problem was then to get him back to camp but we managed this by leaving my motorcycle behind and I rode behind Don holding the pup in my arms. Although the trip back to camp was fairly rough the pup did not seem at all worried, in fact he seemed quite contented. Later that day Don and I doubled back to retrieve the cycle.

It was just before midday when we hit the camp and all the Rebels had returned for the midday meal from various jobs, when we entered the tent they were all bashing the spine (laying down) "What the hell have you got there" exclaimed the Gogg noticing the pup under my arm "Our new mascot" replied Don, the Rebels all got up and came over to my bed where Don and I had sat down, we told them how we happened to find him and they all agreed the pup could do with a good

10

square meal. Fitz first noticed the pup was sporting a few fleas and exclaimed "He's a bit lousy" Murchie piped up with "Well you're a bit crummy (lousy) yourself" "All dogs have fleas" quoted Gordie scratching himself unconsciously. The call of mess parade headquarters from the orderly NCO was the signal for dinner, so we tied the pup to the tent pole and wandered across to mess.

Each of the Rebels returned with a portion of meat for the pup. An old plate was produced from somewhere and a huge feed was placed on the ground within reach of the pup, we retired to our various beds to watch the result.

The pup cautiously sniffed here and there around the plate, then promptly started shovelling sand over the plate, using his nose as a shovel, every now and again he would stop, sniff again, then recommence until he was satisfied that no smell escaped from the little mound. "Wouldn't that rip yer" from Murchie. Gordie said you could scarcely blame him. Fitz reckoned that the meal was on the bugle (nose) alright. The Gogg enquired, "Do you think he may be an Iti dog (Italian) and require olive oil with his food. As the pup was obviously hungry we reckoned it was worthwhile giving the oil a try.

I elected to scrounge a little oil from the R.A.P while Don removed the plate of sand covered meat and washed the sand from it. Lightly covering the meat with olive oil we again offered it to the pup, strange as it seems it was exactly what he required, and to the accompaniment of much tail wagging, he wolfed down the scraps of meat. "How did you work that one out Gogg" inquired Don, "Just fluked it" replied the Gogg. "Righto you blokes on your feet" greeted us as Poppa came into the tent...

Murchie a little reluctantly disposed of his pet asps under the strict supervision of all the Rebels. "You know Cleopatra died by clutching an asp to her breast after learning Mark Antony had been killed", he told us, previous to this we had accepted his word that asps were really harmless. "Murchie will be the cause of white flowers under someone's photo one of these days" declared Poppa, "and if you ever bring another pet snake within my sight it'll be your own photo."

"Well I guess we had better give the little pup a bath" I suggested. Don got a tin and some water from the cookhouse while I scrounged through Feathers gear for his life-buoy soap, the rest of the Rebels disappeared as Poppa mentioned something about a working party for a job. The little pup looked very miserable while we washed him, he certainly had his full quota of fleas, however he accepted the bath in good grace. After the wash we dried him on Fitz's towel, when the toilet was completed the pup was as pleased as punch, he scampered in and out of the tent and up and down upon the beds, making himself thoroughly at home. "Looks like he will stay" remarked Don as we stood watching the happy pup, there was little doubt about it that the pup was prepared to accept our offer to join the 2nd Machine Gun boys, for better or worse.

Later that afternoon when the rest of the Rebels drifted back, the pup, bright and shiny as a new pin, greeted each as they came into the tent with a wag of approval. Shortly before tea Big Jim popped in and seeing the pup picked the willing little chap up and patted him. "What name this fella George" he inquired immediately setting us another problem.

Feathers suggested Longfella, this described the pup alright but the Gogg pointed out that you cannot walk about calling out Longfella, Longfella, if you required the pup, as the rest of the mob would think you were 'desert happy.' Gordie said don't call him George or we will have half the wogs in Egypt answering the call.

Don suggested we call him Roy after our good friend Poppa, but Murchie reckoned it wasn't a fair go for the pup as he looked far too intelligent to be called after Poppa. However Big Jim remarked that as we already had one mascot called 'Erb' why not call the pup 'Orrie' 'Erb and Orrie' usually go together.

We debated this title and eventually agreed to give the pup the full title 'Horace' however as time went on the name developed into Horrie, probably because it was easier to call 'Horrie' than 'Horace'

The question of breed came up also that night. Murchie suggested he might be a cross between an 'Upsetter' and a 'Disappointer'. The Gogg said that as the olive oil strongly suggested some breed of Italian dog he may be an 'Iti-dale'. Gordie declared that he was probably some breed of Arab dog who had once been owned by an Italian, perhaps an Arab terrier. Fitz immediately popped up with 'Arab dog' 'Wog dog' this name caught on all over the camp, if anyone spoke to the dog it was either 'Horrie' or 'Wog dog', but if talking about him it was 'Horrie the Wog Dog'.

Proof of Horrie's belief in protecting the camp against intruders.

Top: Horrie guarding the camp.
Bottom: An idle afternoon with his mates in the Middle East.

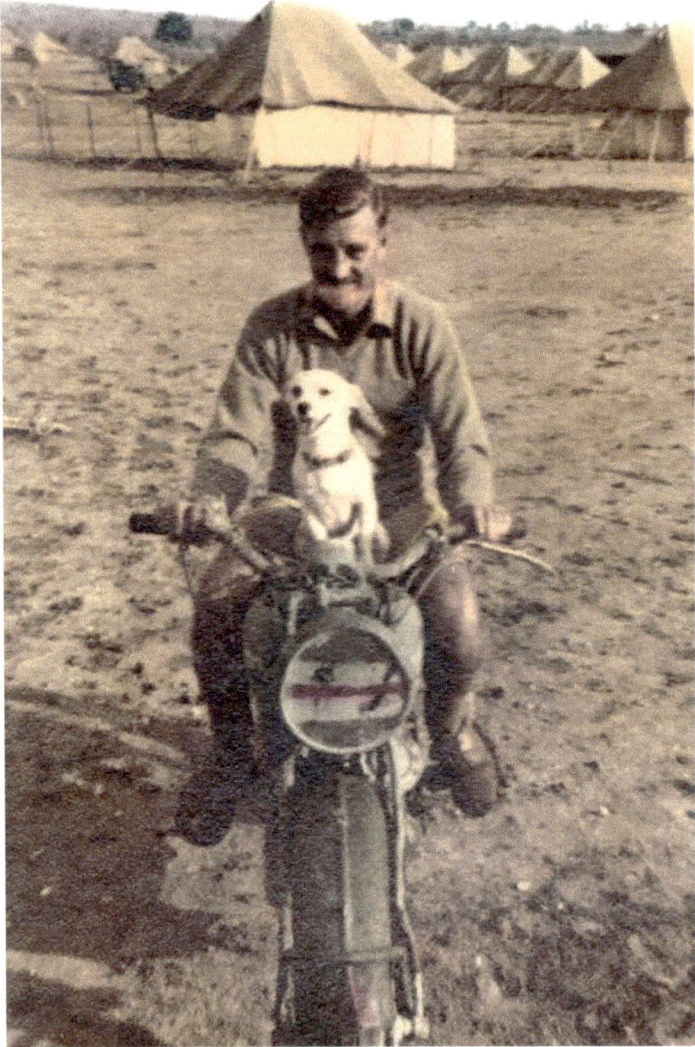

Horrie masters the Company's motor-bike.

The advent of 'Horrie' the little wog dog, he is having a bath a few days before we left Egypt for Greece.

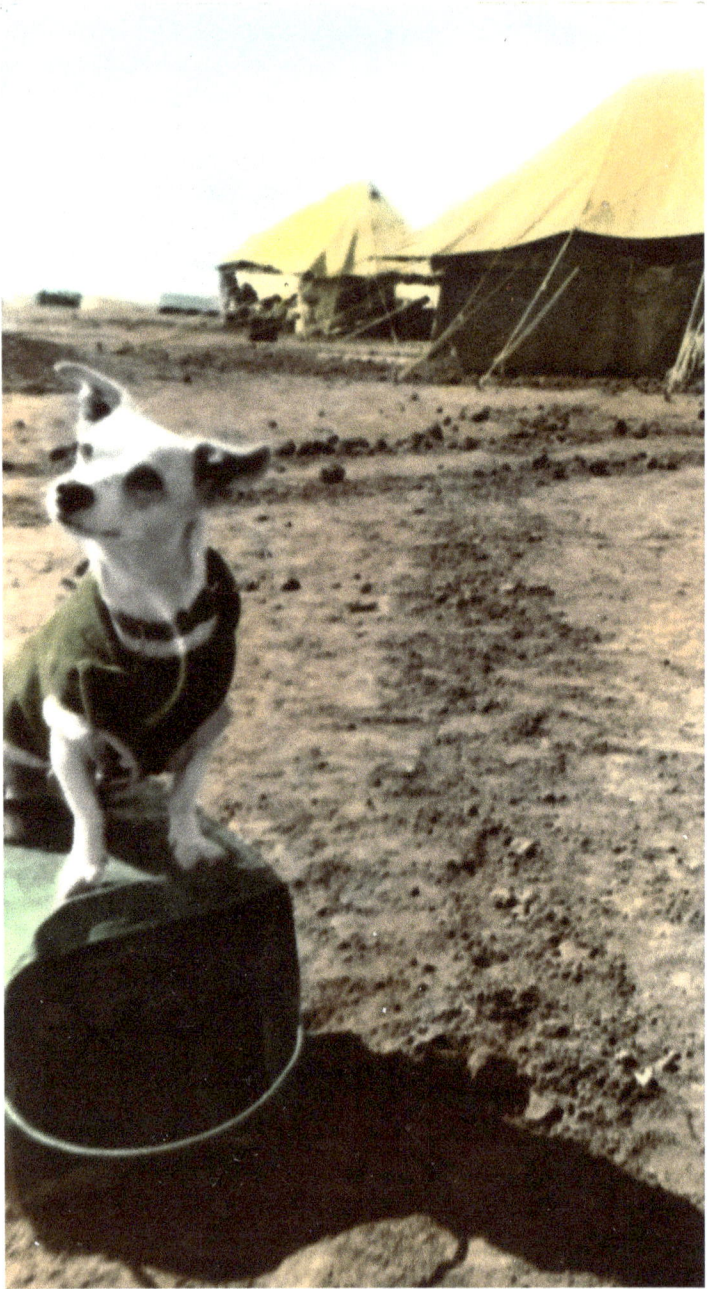

Moody's first photograph of Corporal Horrie in full dress uniform.

December 1941. I wonder if one could see a sight like this
in any army but the Australian. 'I doubt it.'

Corporal Horrie in full uniform, memorising the morse code message.

"Some of the Sigs. and Horrie all wearing the caps that I received in Kay Krugers parcel."

Palestine 1941. Corporal Horrie (now outranking his master)
accompanies Private Moody before settling the troops. .

Corporal Horrie spotting for the enemy, Machine Gun Company 2/1.

Corporal Horrie, on guard 1942.

Top: Horrie examining a bomb crater close to the camp.
Bottom: Horrie, counting rail freight cars.

A break in patrol.

Top: Horrie in Greece, an amateur archeologist.
Bottom: With Don Gibson and James Moody.

New Army camp, same dog!

Horrie with a local fan in Crete.

Corporal Horrie's receives a local salute..

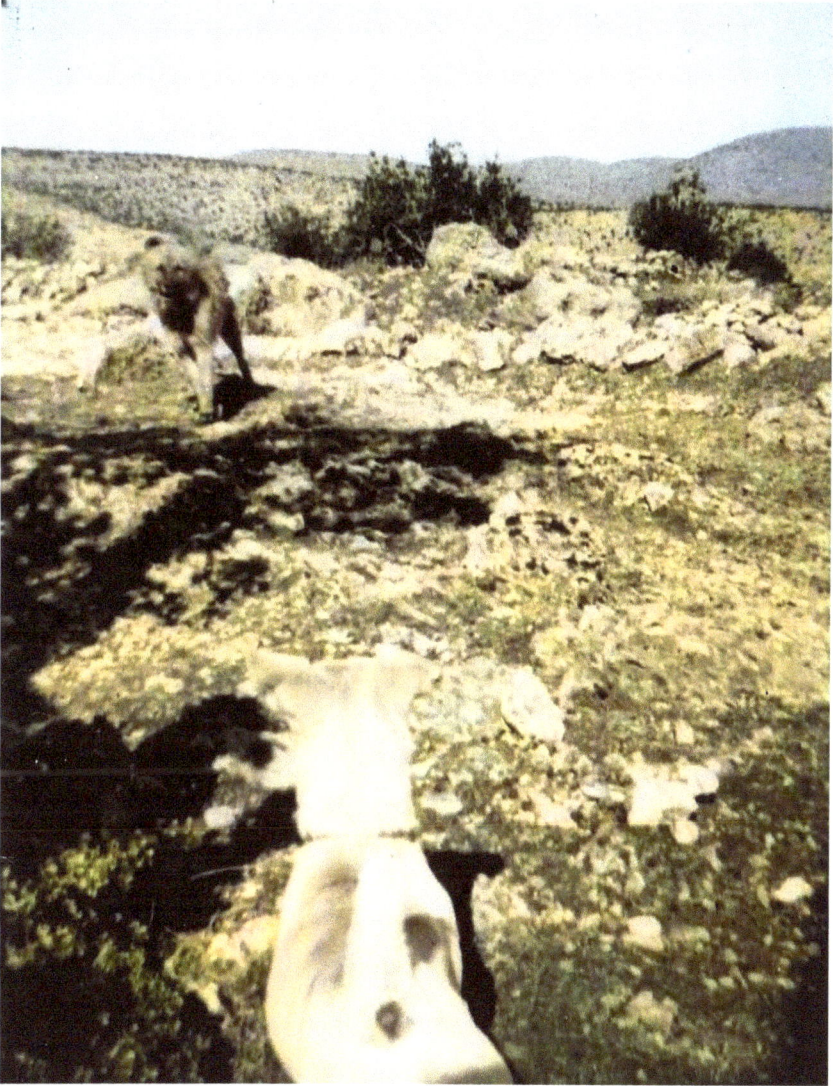

Horrie stands firm against a very scary foe.

Corporal Horrie's vehicle to protect his paws from the deep snow.

Horrie, off the leash.

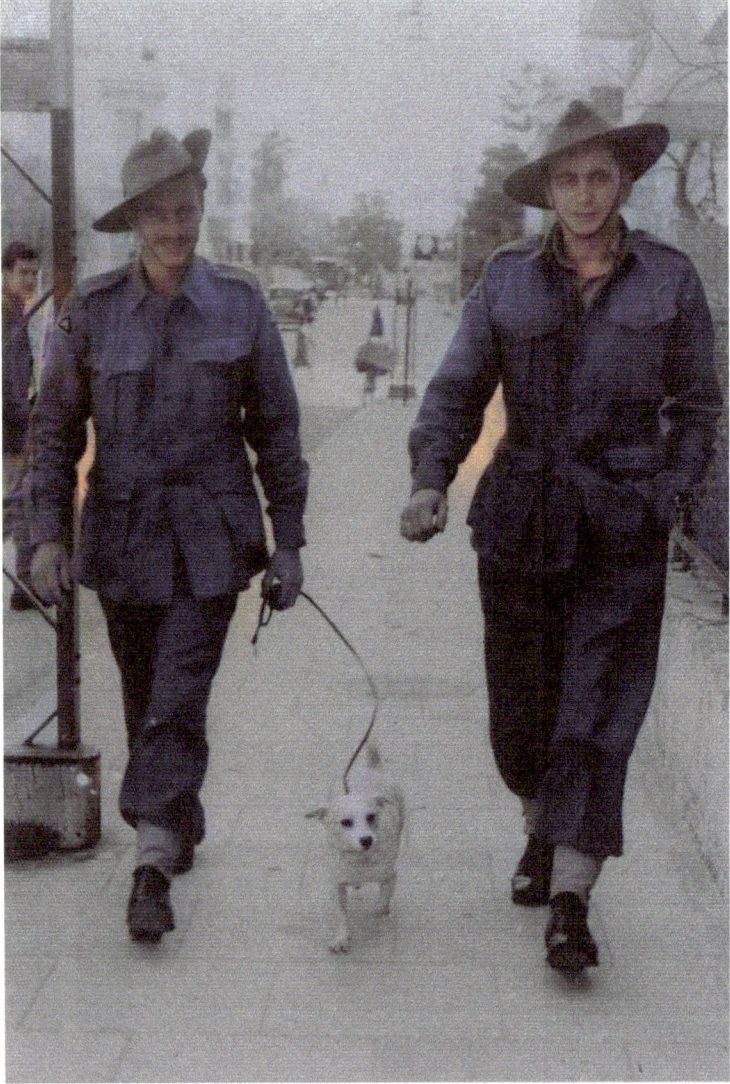

Moody and Gill with Horrie.

Waiting for a beer, including Horrie, 1942.

Top: Horrie leads Moody astray, while "on leave" in 1942.
Bottom: Horrie finds a very large fossilized bone.

Horrie iwith his girlfriend Imshi. Two brave mascots.

Corporal Horrie, Mascot of the 2/1 Machine Gun Company.

Horrie is surrounded by his crew after their freighter went down.

Horrie in Moody's arms, leaving Crete with the Machine Gun unit.

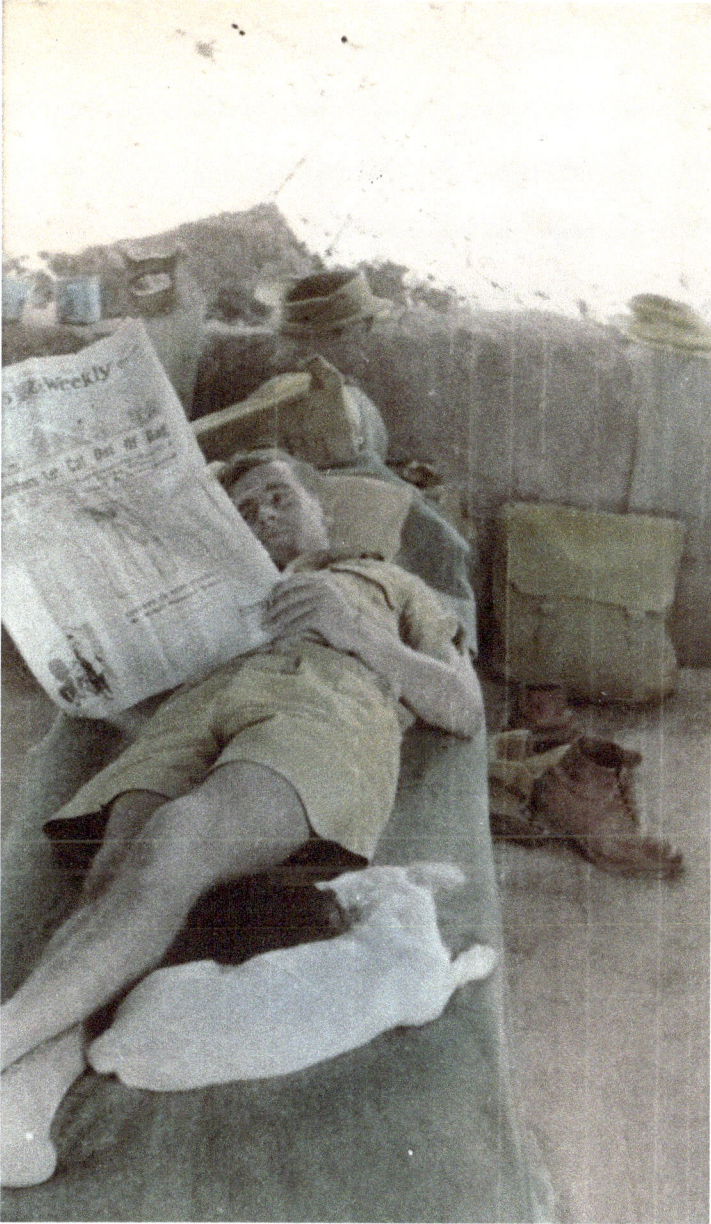

Almost home.

Corporal Horrie being transported back to Australia via the knapsack of
a member of the 2/1 Machine Gun Battallion, in March 1942.

Horrie takes a breather from his back-pack after embarkation.

Moody's modified back-pack remains at the Australian War Memorial.

Headquarters

2/7 AUST INF BN
11 Jun 43

Dear MOODY,

Yes. DIXON is still Adjt of the BN and he does recall having two stowaways on his hands for disposal.

However, we would be most happy to have you back with the unit- this time for keeps- only wish you could have been with us during the flap. I'm sure you would have enjoyed it.

The posn, at present, how we can assist you in effecting the transfer. I'll enclose an official letter with this, stating that transfer to this Bn is acceptable, providing it is concurred with by the unit. Should anything further be required of us, just drop a line and advise what other action you would like us to take.

Regards to yourself and GILL,

Sincerely Yours,

Reginald Dixon.

Australian **Military Forces**

TELEPHONE

Subject: TRANSFER

N/398

Address 2/7 AUST INF BN
11 Jun 43

VX 13091
PTE J.B. MOODY
2/1 MG BN

Ref your application for transfer and that of Pte GILL.

2. It is advised that application will have to be submitted through your parent unit.

3. Should transfer be concurred by them this letter can be shown as proof of our willingness to accept you both.

..............................Major

ADM COMD 2/7 AUST INF BN

After the excitement of war in the Middle East, Horrie sought a transfer to the new battlefields of New Guinea, but missed the action while Moody and Gill stowed away to Rabaul.

Jim Moody and his special mate were two real good blokes. A bit wild, but great company. J.L.I.

J.B. Moody,
Private 2/1st Machine,
Gun Battalion, A.I.F.

Dear Mr Moody,

Mr Cousins, of Angus & Robertson Ltd., has handed me your M.S. re the Wog-dog, with the idia we both may make a book of it.

There undoubtedly is partial material for a good dog book in the M.S. Unfortunately, the material is not sufficient for a real book. The very least number of words a book should run to, is 60,000; really there should be no less than 70,000. Your M.S. runs to approximately 18,000 words. So that our problem is to find material for a further 40,000 concentrated words at least.

I have endeavoured to solve this by writing you a list of questions. Please read these over carefully first, then put the old memory back. Then answer each question (scribble it out in lead pencil) fully as you can. By the time you've got to the end you'll be surprised at how much more you know of Horrie than you previously thought you did. Then go over the questions again, and write the answers out fully for me.

If your mates are still with you, discuss the questions with them. A lot of memories are better than one. You'll find you'll be given a wealth of material to write down. Some of the lads also, will know stories of Horrie that have escaped your notice, or memory.

Any question that brings up some other incident not mentioned in the M.S., then write down that incident also. In this way I'm quite sure we'll get enough material for what should prove a jolly good book. Don't be afraid of answering any questions very fully, or of adding more copy, so long as you send along "meat".

Cheerio. And here's wishing the forthcoming book every success.

Sincerely,

This is how the Wog-Dog was written. I used to send Moody & his mates lists of questions while they were in New Guinea too. They were good blokes. The Wog-Dog went. They were very well too brought in 3 or 4 but-ones percopy 4 me!

12.9.1971

Jon L. Idriess

Idriess first wrote to Moody in June 1943, encouraging him to extend his memoir. His later comments are from his Archive, written in 1971.

QUESTIONS.

A. Please describe, in all detail that you remember, the dog when he first joined up. Colour, size, anything at all about him. And, as he later developed any particular characteristics whatever. As described now, he is but a "shadow dog". Hence, it is very neces ary that any habit, any peculariaty, any characteristics should be described, both on adoption and as he grew, so that a picture of a "live" dog can be quickly built up in the readers mind.

I understand that the dog may be in Sydney. If so, I'd like the address and a chance to go out and see him.

Also, re photos. Any photos of him, together with interesting photos of the Battalions adventures, would help the book immensely. *P.S. Mr Cowell has since told me he already has the photos.*

B. Who is Big Jim Hewitt? He is only mentioned in the very last line, yet apparently he had quite a lot to do with the dog.

As you read the questions you will repeatedly see "give names of cobbers, scraps of conversation etc. This will all help to build up the book. Not only the dog, but the men closely associated with him must be "living characters" for the book to live.

1. Was it on an expedition to the old Roman city that you found Horrie? What were the names of your companions? Do you remember any scraps of conversation re the dog? For instance, who suggested name of Wog-dog? Any particular reason for the name Horrie?

2. Dog covered his feed with sand. Who suggested "you can scarcely blame him"? (names and scraps of conversation always make an incident more interesting) Who was the spark suggested he might be an Italian dog?

3. Who were your tent mates who took most int rest in the dog?

4. While on route marches, the Arab urchins back chat, especially in regard to the dog would win a laugh from the reader. Slip in as many witty remarks or incidents, occurring at any time, or on any occasion, as you can possibly remember.

5. When about to move from Egypt, the question was "What about Horrie." Please think up a few scraps of conversation, any suggestions the boys made etc. It will make the story more personal and convincing. Give names whenever you remember. Also any little details in training Horrie to the kit bag.

6. Any amusing incidents during the train trip?

7. "One of the boys called Yow", to what does this refer?

8. Re ships crew and their dog. A few personal incidents about members of the crew, their dog, and still more of Horrie, aboard ship would add to interest.

9. The sea sickness incident is quite good. Any more cheery little touches would be appreciated — by the reader anyway.

10. When you made a life jacket for Horrie, what were names of cobbers who lent a hand, or took an interest? Throughout the entire book please add these

Some of the 63 questions asked of Moody as they tried to develop the book.

Pte J Moody
"C" Coy.
2/1 M/G Bn A.I.F. 27 July.

Dear Mr Idriess

I am forwarding the enclosed .24 pages
on the installment plan, it being rather
inconvenient for me to keep it until I
have answered all your questions. As you
know this Country you will appreciate
the difficulty in keeping things dry.
I will continue with the remaining quest
ions and hope to get time to finish
them very soon. Would appreciate ack
on receipt of 124 pages, also if any meat
is obtainable from same.

Sincerely   J B Moody.

PS. Paper of any kind is as scarce
as my correctly spelt words up here.

Moody responded, with 124 pages to Idriess in July 1943.

47

(PRO FORMA D.2)

AUSTRALIAN MILITARY FORCES

NE2.JW.GM   "A.R." POST.

         (21.15.9841)

N.S.W. Echelon & Records,
BROADWAY.   SYDNEY.

MEMO TO:

23 JAN 1945

VX13091,
Pte. Moody J.B.,
23 Silver St.,
ST. PETERS, N.S.W.

     Enclosed are the undermentioned in respect of your discharge from the A.M.F. which has been effected as from **4th February, 1945.**

        1. Discharge Certificate No. **124569**
        2. Active Service Badge No. **A.59036**
        3. ~~Army Form A.12 purporting to Contain your Will.~~
        4. Advice re Malaria, etc.

     Please acknowledge receipt of the abovementioned by signing and returning the attached duplicate of this memo.

     Please note that Discharge Certificate is to be signed by you where indicated on the back thereof.

     If you have not already done so, you are to report to the National Service Officer in the Area in which you reside.

                               Stephen
                                   Lt.-Col.

3 FEB 1945   Encl./ **4.**   Officer in charge N.S.W. Echelon and Records.

RECEIPT IS ACKNOWLEDGED
OF THE ABOVEMENTIONED

J.B. Moody.

48

From James Moody's War Record, we can see that Corporal
Horrie served in the Middle East,; and was thus eligible for the
1939-1945 Star, the Africa Star, the Defence Medal, the War Medal
and the Australian Service Medal.

Corporal Horrie with his knapsack, in the Moody backyard
- 28 Meadow Street, St Kilda, February 1945.
*Horrie, the civilian, taken in Melbourne, three years to the month
after we first found him. Horrie is unable to soldier on now, being too
plump for his uniform and pack.*

HORRIE embarks ON LIFE OF PEACE

Seasoned veteran of five campaigns of this war, Horrie the Wog Dog, mascot of the 2/1st Machine Gun Battalion, AIF, has received his honorable discharge.

Horrie, who belongs to Mr. Jim Moody, late of the battalion, was found wandering on the fringe of the Western Desert, Egypt, as a puppy nearly five years ago. He has been with the battalion ever since, Horrie went through Egypt, Greece, Crete, Palestine and Syria, and at one stage was the proud possessor of two stripes, sewn on to his flannel jacket. In Syria, however, he and his owner, then a lance-corporal, went AWL for a few days, and on returning both lost their stripes.

War Wound

In the evacuation of Greece, Horrie was on the Costa Rica, which was bombed and sunk in the harbor. He was thrown from the deck of the ship to a destroyer close by, and caught by Tommy sailors. Wounded in the leg with a shell splinter during the evacuation of Crete, Horrie was operated on with a jack-knife by members of the battalion. He is now in fine shape, and his war injuries have left little mark. During the campaigns he wore an Identification disc made from a Greek coin. It bore the words, EX No. 1, the E standing for Egypt. Horrie's owner is anxious to find another Wog Dog as a mate for. Horrie. In the meantime. Horrie, who will be made an honorary member of the RSL, Melbourne, has decided to offer his services to the Red Cross to raise money.

Sun (Sydney), Monday 12 February 1945, page 3

Corporal Horrie who has been in several campaigns, received
Honorary Membership of the Melbourne RSL as published on
12 February 1945, prompting the Department of Health to
impound our hero.

18 Meadow St
East St Kilda
28th Feb.

Dear Mr Idriess

Received your letter of the 7th Feb
glad you appear to be satisfied with
the first chapter and trust you will
be able to continue the good work
throughout the book. At the
moment am at home on leave, but
I expect to be in Sydney about the
20th March, if there is anything I can
do, or if you so wish it, I will
contact you in Sydney. If you desire
this, lets know before the 18th March,
the above address will find me.

Cheerio

Jim Moody.

Ion Idriess congratulates Corporal Horrie for his war-time service.

28 Silver Street,
ST. PETERS.

2nd March, 1945.

Mr. R. M. Wardle,
Director,
Division of Veterinary Hygiene,
Department of Health,
CANBERRA, A. C. T.

Dear Sir,

In reference to what your department probably considers
to be an alien and prohibited dog, I am making this direct appeal
to you, trusting you will accept this epistle in the spirit in
which it is penned.

Firstly, I accept full responsibility for smuggling the dog
into Australia and I am quite prepared to face up to any charge
your department may prefer against me. It is not my intention to
avoid the issue, but rather accept the penalty with the least
possible trouble to the department concerned. With this in mind,
I am placing myself quite voluntarily at your disposal, but I do
wish to appeal to you to give the dog your every consideration.
I state the following facts that you will, I trust, consider in
your judgment.

The dog, now some five years of age, was found by myself
when he was quite a pup wandering on the fringe of the Western
Desert. The pup was adopted by the Signal Platoon of the 2/1
M.G. Bn., and he remained with my unit for a period of about 18
months, during which time we served in Egypt, Greece, Crete,
Palestine and Syria. Perhaps you will understand that there
existed a very strong affection for the dog by members of my
unit after this service. The dog was well cared for, and he
never at any time showed signs of any cannine disease. Knowing
the risk of rabies in the Middle East, I took the dog to a
Vet. in Tel Aviv just prior to embarkation to return to Australia.
The dog was carefully examined and pronounced free from any
disease. With this knowledge I smuggled the dog home, hoping to
have him quarantined after arrival in Australia. There was no-
one in authority aware of this episode.

After arriving in Australia it became apparent that should

Moody takes responsibility, hoping to save his dog.

I declare the dog he would be destroyed and not quarantined
as I had hoped. Consequently I kept his presence in Australia
a secret from the time of his arrival, April 1st, 1942, now
almost three years ago.

The dog's service with my unit was of such interest that
Ion L. Idriess has written a book about him. The book is to
be published this month. I realised that with the publicity the
dog will receive through the book, he could be made an asset to
the Red Cross in their appeal for funds, so I offered the service
of the dog to them. They were in complete agreement that the
dog would be of valuable assistance to them, and were pleased to
accept my offer. The first job for him is being an attraction at
the 1945 Easter Show to be held at the Lady Gowrie Red Cross Home,
Gordon. The dog will no doubt be of great assistance in raising
funds at various patriotic functions. With this in mind I have
brought him to the public eye through the newspapers and have
consequently subjected myself to the penalty of my misdemeanour
in smuggling the dog into Australia.

I do not wish to appear flagrant, but I would like to
point out that it would be a comparatively easy matter for me
to have kept the dog's presence in Australia a secret, but by
doing so I would be denying the Red Cross of a helpful servant.
To do this I am prepared to accept the penalty, but I wish to
appeal to you to spare the needless destruction of a healthy
and faithful dog that can be used for such a worthy cause as
the Red Cross appeals.

I feel sure this appeal will meet with the fair and just
British spirit in which I believe and have volunteered to fight
for during the past five years.

Thanking you,

I remain,

Yours faithfully,

J. B. Moody

TELEPHONES: MA6511 (5 LINES)
CABLES & TELEGRAMS-"FRAGMENT" SYDNEY

Angus & Robertson Limited
ESTABLISHED 1886

PUBLISHERS   BOOKSELLERS (NEW & SECONDHAND) & LIBRARIANS

"THE SYDNEY BOOK CLUB"
AUSTRALIA'S
LEADING CIRCULATING LIBRARY
CATALOGUE & TERMS
ON APPLICATION

89-95 CASTLEREAGH STREET, SYDNEY.
BOX 1516 D.D. G.P.O. SYDNEY.

ALL COMMUNICATIONS TO BE
ADDRESSED TO THE COMPANY

2nd March, 1945.

Mr. R. M. Wardle,
Director,
Division of Veterinary Hygiene,
Department of Health,
CANBERRA, A.C.T.

Dear Sir,

We are to publish a book shortly entitled "Horrie the Wog Dog". This book is written by a soldier named Moody who brought it to us in diary form. We then got Mr. Ion L. Idriess to write up the story from the diary. Enclosed are galley proofs which you might like to look over. The story is undoubtedly the best dog story of the war.

Unfortunately the diarist did not, I find, get in touch with the Quarantine Department when the dog was brought into Australia, being terrified that it would have been destroyed.

The dog went through the Greece and Crete campaigns with the soldiers and the two ships he travelled on to Greece and back from Crete were both torpedoed, but the dog's life was saved. I understand the dog had been three years in Australia and do hope that his life may be spared.

Yours faithfully,
ANGUS & ROBERTSON LIMITED.

W. G. Cousins.

MAR. 5. 1945

Angus & Robertson sent Mr Wardle the galley proofs of Horrie's story.

V213

Dear Sir,

   I am in receipt of your letter of the 2nd March accompanying proofs of the book "The Dog Dog".

   I am afraid I fail to appreciate the story and surprise is expressed that your firm would countenance a publication which records a deliberate breach of the law.

   The dog has been formally taken over by our officers and, by direction, it has been destroyed.

   It might interest you to know that the Middle East, and in fact almost every country in which the present war has been and is still being waged, is rife with Rabies.

   Australia is the one large country in the world which has been kept free of this dread disease of animals and man and our Animal Quarantine Service is carrying the grave responsibility of endeavouring by all possible means to keep it so.

   It is to be hoped that the story will never have to be written, recording how some months after the smuggling into Australia of a dog from overseas, the disease of Rabies occurred, and in which would be described the suffering and mortality of man and animals concerned in the outbreak.

   In the book "San Michele", Chapter V gives a vivid description of Rabies in human beings, and of the psychological reaction in man in coutries where rabies is endemic, to the effect of dog-bite without knowledge whether the attack is made by an infected dog or not.

                              Yours faithfully,

                              C.F. ARDLI.
                              Director, Division of Veterinary
                              Hygiene.

Messrs. Angus & Robertson Ltd.,
Box 1516 D.D., G.P.O.,
SYDNEY.   N.S.W.

The Department's response was swift, and vicious.

Moody was forced to pass over the dog to the Abbotsford Quarantine
Centre on March 6 and telegrammed the Director of Veterinary
Hygiene, Department of Health, Mr Wardle who responded abruptly:
"... the dog is to be disposed of by destruction..."

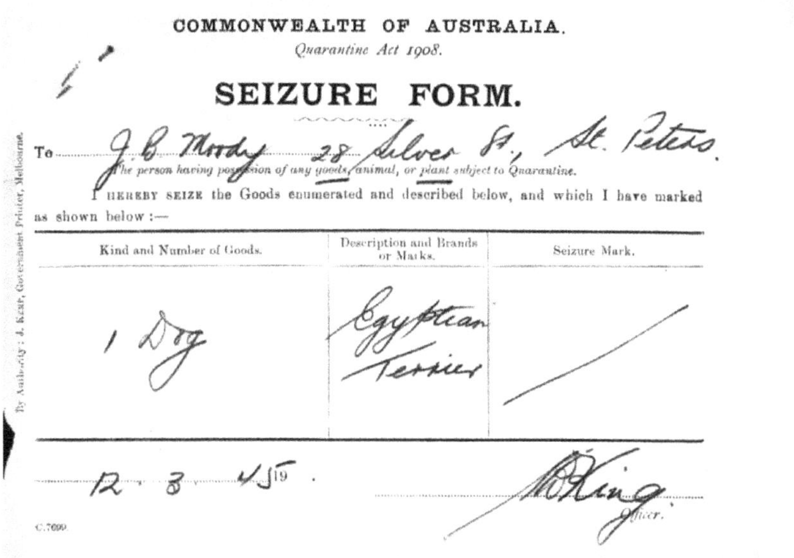

COMMONWEALTH OF AUSTRALIA.

Quarantine Act 1908.

SEIZURE FORM.

To _J. B. Moody_ 28 _Silver St, St. Peters._

The person having possession of any goods, animal, or plant subject to Quarantine.

I HEREBY SEIZE the Goods enumerated and described below, and which I have marked
as shown below :—

Kind and Number of Goods.	Description and Brands or Marks	Seizure Mark.
1 Dog	Egyptian Terrier	

12 · 3 · 45

M. King
Officer.

C.7600

By Authority: J. Kemp, Government Printer, Melbourne.

After several years of service in the Australian Army, and almost three
years of a quiet life at Moody's family home in St Kilda; Horrie was
proclaimed a Dog to be Destroyed on March 12 1945 and dispatched
at 4pm. Moody was given 5 minutes notice.

(To The LowesT cLass of people_ I ever
Heardw of) who ordered a faithful
Dog to be puT To deaTh To geT
their own back because The young
soldier responsiblee did noT go
gushing all over you. we The public
of AusTraLia are Led To belieue ThaT
iT was To puT down dicTaTers, bruTaLiTy,
and Terrorism, and build a worLd of
peace and Love. ThaT ouR dear Boys
are away fighTing These braue Lads
had Love To spare for a poor LiTTLe
pup in face of alL their own dANgers
thought they were bringing him home
To This fair LANd we all boasT of To the
ouTside world. where They would be
weTcomed wiTh open arms, iNsTead
nothing buT dicTaTers, cowards, and
bruTaLiTy greeTed them. iTseemsA piTy
The coward ThaT shoT The dog did noT
puT his caRTridges To beTTer use the_
new world wonT have A pLace for people
of your Type. I donT know why your wives
would eaT a morseL of food ThaT your
dirTy filThy wages boughT
      may God bless The dear boy in the
Loss of a faithfuL friend (soldier's hiother)
   Horrie has found his new world

The Department received an immediate response from the public.

522 Parramatta Rd
Petersham
17/3/45

Mr Jim Moody
Dear Sir

Having read of your sad loss in yesterday's Truth, of the brutal killing of your mascot, poor dog, he was a hero, & you should be decorated for your kindness to dumb animals instead of being penalised. As an Animal lover & member of the R.S.P.C.A. Would you kindly ring up L. M 1693 Tuesday or Wednesday night, we are going to see what can be done in the matter. Please ring after 6. pm. L. M 1693.

Yours faithfully
Mrs C. M. Clark

41a, Surrey St,
King's Cross,
Sydney.
April 23rd/45.

My dear Jim Moody.

When I read the account of Horrie in the "Truth" of March18th, it made by blood boil to think that it could happen at all. For the remainder of the day I could not read my paper, or settle down to anything, but on the next morning I settled down to put into verse what "Truth" had printed, it was my intention to come out to bring it to you on the Monday night, but I was too sick, and I have been waiting till I felt well enough, so I am posting it to you. I am indeed very sorry that I could not be at the Town hall on Thursday night, as I think that all present would have liked to hear these few verses that I have wrote about Horrie. You see Jim I am an animal lover, and I was in the 1914-1918 war, so I know what a digger thinks about his pal, human or animal, so I know how it must have hurt you, and I trust that something will be done to prevent our diggers pets from being destroyed in the future by such an inhuman devil as was responsible for the death of Horrie.

Yours for a fair go

Tom Brent

To Mr Jim Moody.
28, Silver St,
St Peters.
Sydney.
N.S.W.

61

Sir

    I trust your manly soul - if indeed you possess
a glimmering of such a thing -is now satisfied. What a
HERO you must feel in having rid this small country of
such a huge ferocious wild animal - such a menace to
public safety - you, who, unlike the brave little fellow
you have done to death, probably have never smelt the
powder of a battlefield ! ! May God, Who is the God of
animals as well as humans; and Who, probably, is far more
proud of His quadrupeds than of his bipeds, reward you
as you have rewarded the loyalty, love and devotion of a
small dog. I, personally, regard your brave action as
beneath the contempt of all decent people.

               yours contemptuously

               aj Keelan

                            "Otuaroa"

                          Thirroul. 19.35

Dear Mr Moody
        Yesterday I cut from "Truth" the page about
your lovely little dog, wrote this note, and affixed it
to the page. Then I posted it to

      The Director General of Vetrinary Hygeine

        Canberra.

                    (over)

● St. Bumble And The Wog Dog

Dedicated to the brave Bumbles of the Federal Quarantine Department, who fearlessly executed Horrie the Wog Dog, mascot of 2/1 MG Battalion, AIF, after three years' faithful service in his adopted country.

A dog has bureaucratic ways,
He likes to boss the show—
As every rat and alley cat
In town soon learns to know—
But a dog can still be popular
Where a bureaucrat is not—
A truth to smart in the jealous heart,
And put the pup on the spot. . . .

Also the dog has four good legs
To his human rival's two—
Which is contrary to the majesty
Of Red Tape Ballyhoo;
But a dog who's been in Greece and Crete
Is one to be feared and hated—
Above his station—By Regulation
He's gotta be Liquidated!

Following Horrie's demise, Moody fights back, getting his photographs of Horrie to Idriess in this letter of March 20 1945, while Idriess rewrites the ending of their book.

28 Meadow St
East St Kilda
Melb 20th 45

Dear Mr Idriess

Your letter to hand today, note your request for snaps. I will be [in] Sydney about 5th April and will bring all the snaps that I have of Horrie. The snaps I have of Horrie are not enlarged like the previous lot you had, consequently they are a little clearer and may be more suitable, anyway I will bring all for you to select from. Glad you seem to like the book and feel confident it will be well received.

Cheerio for the moment

Jim Moody

61 Cambridge Street,

Paddington.

22nd March 1945.

The Editor,

"Truth,"

Dear Sir,

"Horrie."

This is just to express my horror at the needless and callous killing of Horrie! But the execration felt on all sides at the destestable action of an official will not bring the little canine hero back to life nor assuage the feelings of grief and anger felt by his soldier friends. The said official cannot have much love or respect for our soldiers or he would not inflict such an insult and such an injury on them. "Love me, love my dog."

Of all the weak excuses for a vile action one of the weakest was the excuse recorded in Tuesday's Mirror.

Could there not be a monument erected to Horrie and his history given with full particulars as to his death?

Yours faithfully,

(Miss)   M. Collis

Just one of hundreds of letters received by The Truth in response to the seizure of Horrie.

Flinders
Victoria
26ᵗ 3. 45

Mʳ J. Moodie
28 Silver St
Sᵗ Peters.

Dear Sir,
I, with many of my friends
were astounded at the despicable action
of the Quarantine authorities, at the
legalised murder, of your little dog.
published in the Truth of March 24ᵗ.
As a dog lover, and the owner
of an Aust. terrier, now ten years old,
I extend my deepest sympathy in
your loss, and can only regret, with
you that someone could not be punished,
for such a dastardly act.
Yours sincerely
K. J. Curtis.

World League for Protection of Animals

(AUSTRALIAN BRANCH) Reg. No. 2835.

and the

Australian Humane Education Society

Including HAPPINESS FOR ANIMALS' CLUB (Youth   a).

Patron—Mr. J. C. Bendrodt.
President—Mr. A. Tonge, M.L.A.
Deputy President and Treasurer—
Mr. J. B. Steel. Phone, MA 3111.
192 Castlereagh Street
Sydney.

Hon. Secretary—
Miss E. B. Moore. Phone, LL 2175
41 Pine Street, Marrickville.

MEETING P   E AND LIBRARY:
Chartres House, 1st Floor
309 George Street, Sydney
(Next Wynyard Station)

Organ of the League-Society—
"The Animals Champion
of Australasia"
(2/6 posted)

April 17th, 1945

Mr Ion L. Idriess,
Sydney.

Dear Sir,

Would it be possible for you to attend the Public Protest Meeting, on Thursday evening next, 8pm, at The Town Hall, Sydney (Basement) to express indignation at the killing of "Horrie, the Wog Dog" ?

Mr J. Moody, the owner of the dog, attended a joint meeting of the above League-Society, and the Howard Prison Reform League, at which arrangements were made to hold the Public Meeting on the 19th April. Mr Moody informed us that you were publishing a book in connection with the whole story about "Horrie", and that he hoped you would be able to be present at the meeting, and be one of the speakers.

It is proposed to take political action to ensure that such slaughter will not happen again.

I enclose, with compliments, a copy of our small quarterly magazine.

Yours Faithfully,

Evangeline B. Moore Hon. Sec.

Our Aim: To promote Animal Welfare and Animals' Rights in every way possible.

"Indignation at the killing of Horrie the Wog Dog" - April 1945.

"Ettrick",
849 New South Head Road,
Rose Bay.
N.S.W.

3rd July 1945.

Dear Mr Moody,

I have just finished reading "Horrie" the Wog Dog.

Having read numbers of Dog Stories I think I am priviledged to say that "Horrie" is and always will remain the greatest ever written.

Mr Idriss has certainly captured the true spirit of your Diary and "Horrie" and his noble band of Rebels live and breathe in every page.

I found it quite impossible to read the touching little Epitaph dry eyed, and one wonders how any human being could have brought themselves to take on the job of destroying such a wonderful little fellow who apparently was in perfect health up to the last.

I consider that the story of "Horrie" along with his coat (so professionally made) and his Travelling pack should have a place of Honor in Australia's War Museum at Canberra.

The story of "Horrie" amd his wonderful comrades, not forgetting "Imshi" the vamp, will live forever.

I like to think that one day we will see his story transferred to the screen, and why not; "Lassie Come Home was.

The Magician turn put on for Big Jim in W.A. was a highlight of the book.

The best of luck to you and your pals.

Yours sincerely,

Addie M. Scroggie

The cover of the first Australian edition published in June 1945.

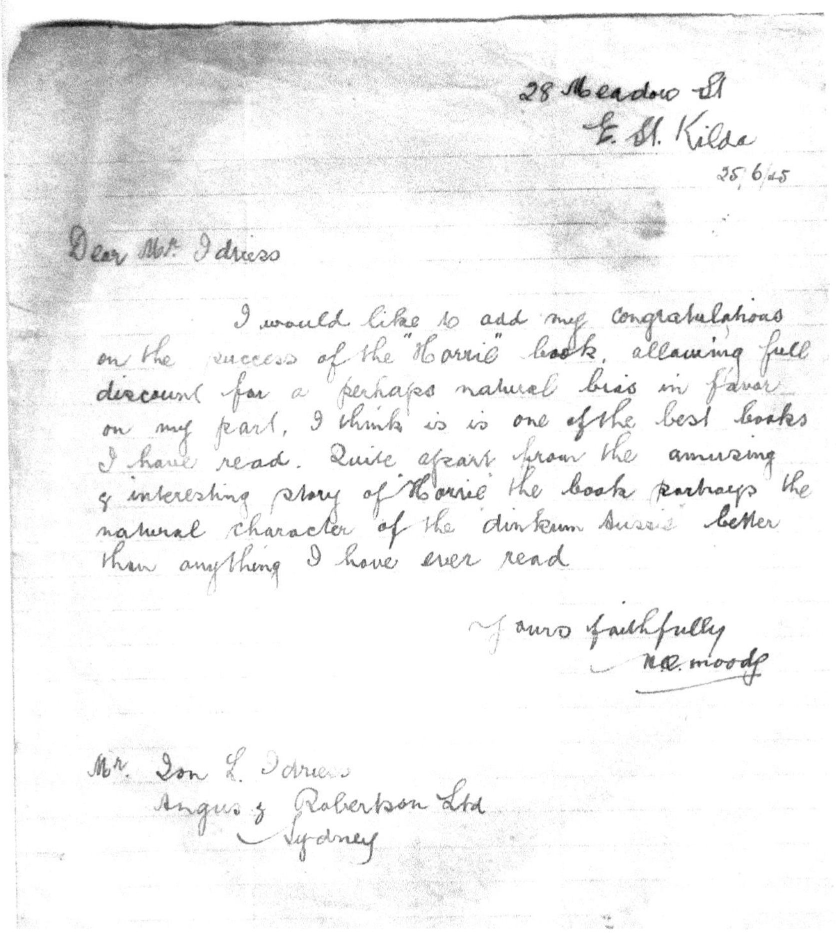

28 Meadow St
E. St. Kilda
25. 6/45

Dear Mr Idriess

I would like to add my congratulations on the success of the "Horrie" book, allowing full discount for a perhaps natural bias in favor on my part, I think is is one of the best books I have read. Quite apart from the amusing & interesting story of Horrie the book perhaps the natural character of the dinkum Aussie better than anything I have ever read

Yours faithfully
N.Q. Moody

Mr Ion L. Idriess
Angus & Robertson Ltd
Sydney

A congratulatory letter to Idriess from Moody's father.

QUARANTINE REGULATIONS

A SOLDIER'S DOG

WAR CRIMINAL?

L Jones.
Powder Works Rd
Narrabeen
Sunday.

To
The Editor
Truth,
Sydney.

May I congratulate
you on your stand in exposeing those
who were responsiable for destroying
the dog of one of our soldiers.
The person who fired that shot
can not claim he carried out his duty,
there is no duty in takeing the life of
any living thing man or dog after
giveing such faithful service to our
country. I would be grateful if you
would find room in your valued paper
to publish the protest of five people
in my home this evening, and convey
our deepest sympathy to the digger

Sincerely
NX 72899.

Ion L. Idriess Esq.,

V

Dear Sir,

I he.ve almost completed reading your account
of Horrie the Wog dog. At the time of his demise -
I should say murder - I was deeply distressed that
such a dastardly act should have been permitted.
It all seemed so very unnecessary. And now, lmowing
the character of the dog, this feeling is ell the
more poignant.

I have always been particularly fond of
dogs, and the fate of Horrie has a nasty habit of
quite frequently cropping up in my mind.

I am anxious, if possible, to have & photo
of him to paste in your book, and am ,.,ondering if
you can tell me how I can go about obteining one,
for whicli, of course, I will pay. I hcve the Sun's
article, and this too is part of the book.

Yours faithfully, ·

(D.) C. R. S. Roberts

1 Glencoe Avenue,
Wellington, C.1.
New Zealand.
28th August, 1945.

Dear Mr. Idriess,

I finished reading "Horrie the Wog Dog" a few days
ago and enjoyed every word of it - expect the last few. I
nearly cried my eyes out to think of that great little dog
having to be destroyed and wished such horrible curses on
whoever was responsible for it that I'm almost afraid they
will rebound on me.

I'm enclosing a letter to Mr. Moody and I would be
very much obliged if you would forward it on to him for me.
It is very much on the above lines. Of course some time has
gone by since the book was published and if, in the meantime,
Mr. Moody has gone the way of Horrie (I sincerely trust he
hasn't) I would be very grateful if you'd read his letter and
perhaps you could answer my questions. I hope I am not putting
you to too much trouble.

Yours sincerely,

(Miss) Grace Burgess

One of hundreds of letters from the public following the release of
the book *Horrie the Wog Dog* in June 1945.

Following publication Ion Idriess received many letters like this: "Could
you please tell me the reason why Horrie had to be destroyed."

A proposed Memorial Fount to Horrie by Mary Edwards, published in the December 1945 issue of *Animals*.

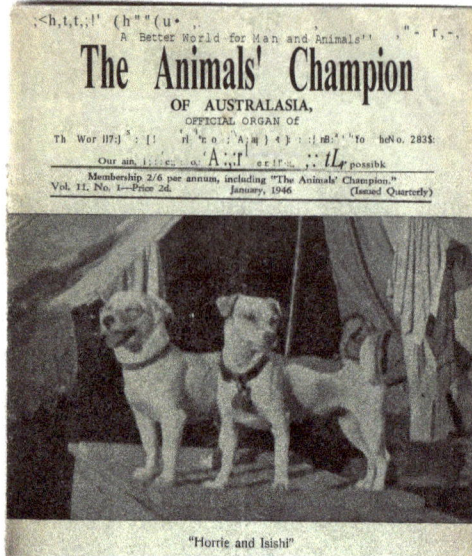

.,<h,t,t,.!' (h""(u• ,.
A Better World for Man and Animals''   ' ' ' r,·,

The Animals' Champion

OF AUSTRALASIA,

OFFICIAL ORGAN Of

Th Wor ll7:]ˢ: [I   rl ʼEo.ʼ A;ʍ } ⸴ }:ꞏ:[ꞏtB:'ꞏ'To   hɛNo. 283$:

Our aim, I:::ꞏ: o.'A:;:l erlf:ꞏ. :ꞏ lL꜀ possibk

Membership 2/6 per annum, including "The Animals' Champion."

Vol. 11. No. 1—Price 2d.   January, 1946   (Issued Quarterly)

"Horrie and Isishi"

HORRIE.   OUR LITTLE CANINE HERO
March 12th, 1945.

We won't forget you, little pal,
Canine "Hero" of our fighting men.
Wounded and sick, you carried on,
Uncomplaining of the deadly bomb.
Unafraid of shot and shell,
You would go through fire and hell,
To fight for the land we're fighting for.
Men who loved you knew your worth;
They smuggled you to "Australia's earth."
Your soldier master brought you home,
To live in peace, to reward you, your bone.
Science said your blood was clean,
(How sad this was unheeded),
Bureaucracy discovered you,
Although your soldier master pleaded.
Unfeelingly, with "Death" they thee rewarded,
Where "Little Pal" the Nation's praise you well-deserved.
In our hearts we mourn your loss,
As we mouth with deep regret
"Our Australian Heroes."
Among those brave "Henoes," little pal,
We will remember y o u -
"We won't forget."
(Put into verse from the cutting, "We won't forget.")
M M English, Voluntary Worker.

76

Horrie's Corporal's uniforms at the Australian War Memorial.

2/1ST AUSTRALIAN MACHINE GUN BATTALION
1939–1946

GREECE 1941	BORNEO
MOUNT OLYMPUS	BALIKPAPAN
SERVIA PASS	MILFORD HIGHWAY
MIDDLE EAST 1941	SOUTH-WEST PACIFIC
CRETE	1945

THE BATTALION ALSO SERVED IN

GREAT BRITAIN 1940   PAPUA AND NEW GUINEA 1942–43

THEY SERVED WELL

WE WILL REMEMBER THEM

A Preface to our Horrie book by Ion Idriess

A Dog that was a better man than me. A tiny dog, so faithfull to his human friends that he never whinged upon the burning sands of the desert, never whimpered when his tongue turned leather under thirst, never moaned when his tiny legs collapsed under him as he struggled on and on and on. A dog who never said "Die!" under shell and bomb and bullet. Who was the liveliest and cheeriest of them all when their ship was sinking under a hostile sea. A dog who barked in a frenzy of defiance at the shrieking wings of stealth that hurled death at them all. A dog who carried out his job and defied machine gun bullets by day, a tiny dog like the wisp of a ghost who defied the barking terrors of night as he sped down rocky mountains with messages of life and death that must be delivered to those waiting below.

I write this book to the memory of a tiny Dog, a wisp of life in a furry coat who never, never let his friends down.

Previously unpublished, from the Idriess Archive

Corporal Horrie's foot prints. .

The Endgame

In 1945 newspapers around Australia carried sympathetic stories and cartoons about the affair, but Moody was ordered to hand Horrie over to the quarantine authorities, who were likely anticipating a welter of such mascot animals to arrive with troops coming home from the war. The dog must be made an example of – particularly if a book was to be published about his exploits in war-time.

IN June 1945, Horrie's story was published by Bobbs-Merril in the USA as *Dog of the Desert*. It was reviewed most favourably but as the war was coming to a close, attitudes to gallant animals had changed - except in Australia. In England mascots were receiving medals for bravery, and in the USA, there was a very different attitude to "the firing squad."

aviation, and information posts.

FULL MILITARY HONOURS AT DOG'S FUNERAL

OUR STAFF CORRESPONDENT.
NEW YORK, March 10.—
Marco, an eight-year-old shepherd dog, who served with the American Army in Africa, will be buried with full military honours at Newark.

There will be a firing squad, an American Legion colour guard, and a flag-draped coffin decorated with the Purple Heart with oak leaf cluster.

Marco suffered one wound in Tunisia, and was twice struck by shell fragments at Anzio beachhead. Complications from a neck wound caused his death.

Mustang Is Fi

Horrie in the kitchen at Corryong, May 1945.

An official letter from the Department of Agriculture to the Director General of Health, Division of Veterinary Hygiene was sent on 14 March 1945 advising that a dog known as 'Horrie the Wog' was destroyed at 4 p.m. on 12 March, under Section 68 of the Quarantine Act. It further states that Moody brought the dog to the Abbotsford Quarantine Station on 9 March. The letter records brief details of Horrie's movements since arriving in Australia. Some reports have stated that the dog was shot at the Quarantine Station but Australian War Memorial records show that he, or his hapless replacement, was destroyed with hydrogen cyanide.

Wait up!

"His hapless replacement"?!

Legend and family memory has it that Moody brought a stay of execution by one day by claiming that he had to retrieve the dog from Melbourne, where Horrie was living with his father. This was permitted, again buying time for Moody to plan a substitution, acquiring a Horrie 'look-alike' from 'the pound' to be destroyed in his place. A dog already on death-row. His Rebel mates smuggled Horrie to a farm in the Corryong District of Northeastern Victoria - a Cudgewa farm run by Eddie Bennetts who had served in the same battalion as Moody.

Moody was still under the threat of gaol for brining Horrie into the country, and naturally was tight-lipped about this while the whole country was on mourning - for all the animals killed at this point in the war.

Family stories from Corryong connect the two families as friends and note the arrival one day in 1945 of a small white terrier that they called "Benji". This was Horrie's home, but it was to be fairly short-lived. The little dog who had survived war was knocked down by a speeding car and died in 1948, but not before siring at least two litters.

Horrie's story inspired Sandra and Richard Hubbard, members of the Corryong Tourist Association to claim Horrie for the area, immortalising him in a bronze statue, taken from Moody's photograph of the dog. It sits in the Memorial Park near the RSL. "Lest We Forget."

Horrie on patrol in the safety of Corryong.

Lee Moody, daughter of Private James Moody, with the statue of
Horrie on the corner of Hanson & Donaldson Streets, Memorial
Gardens, Corryong, Victoria (courtesy of Ian Moody).

Moody's Tail

THE STORY OF
CORPORAL HORRIE

—

JAMES BELL MOODY

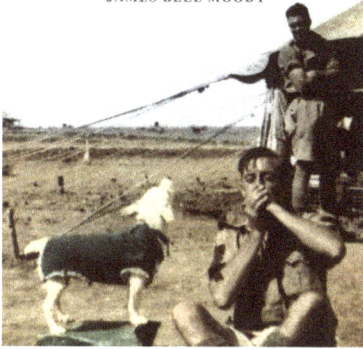

HORRIE THE
WOG-DOG

ION IDRIESS

IMPRINT
CLASSICS

Our sources: Available from ETT Imprint, Exile Bay.

ETT IMPRINT has the following ION IDRIESS books in print in 2025:

Prospecting for Gold (1931)
Lasseter's Last Ride (1931)
Flynn of the Inland (1932)
The Desert Column (1932)
Men of the Jungle (1932)
Drums of Mer (1933)
Gold-Dust and Ashes (1933)
The Yellow Joss (1934)
Man Tracks (1935)
Over the Range (1937)
Forty Fathoms Deep (1937)
Madman's Island (1938)
Headhunters of the Coral Sea (1940)
Lightning Ridge (1940)
Nemarluk (1941)
Shoot to Kill (1942)
Sniping (1942)
Guerrilla Tactics (1942)
Trapping the Jap (1942)
Lurking Death (1942)
The Scout (1943)
Horrie the Wog Dog (1945)
In Crocodile Land (1946)
The Opium Smugglers (1948)
The Wild White Man of Badu (1950)
Outlaws of the Leopolds (1952)
The Red Chief (1953)
The Silver City (1956)
Coral Sea Calling (1957)
Back O' Cairns (1958)
The Wild North (1960)
Tracks of Destiny (1961)
Gouger of the Bulletin (2013)
Ion Idriess: The Last Interview (2020)
Ion Idriess Letters (2023)
Walkabout (2024)

www.ingramcontent.com/pod-product-compliance
Lightning Source LLC
Chambersburg PA
CBHW050819090426
42737CB00021B/3448